Environmental Disasters

By
Arnold Lawson

Environmental Disasters

Author: Arnold Lawson

Copyright © Arnold Lawson (2025)

The right of Arnold Lawson to be identified as author of this work has been asserted by the author in accordance with section 77 and 78 of the Copyright, Designs and Patents Act 1988.

First Published in 2025

ISBN 978-1-83538-828-0 (Paperback)
978-1-83538-829-7 (E-Book)

Book cover design and Book layout by:
    White Magic Studios
    www.whitemagicstudios.co.uk

Published by:
    Maple Publishers
    Fairbourne Drive, Atterbury,
    Milton Keynes,
    MK10 9RG, UK
    www.maplepublishers.com

The views expressed in this work are solely those of the author and do not reflect the opinions of Publishers, and the Publisher hereby disclaims any responsibility for them. This book should not be used as a substitute for the advice of a competent authority, admitted or authorized to advise on the subjects covered.

A CIP catalogue record for this title is available from the British Library.

All rights reserved. No part of this book may be reproduced or translated by any form or by any means, electronic or mechanical, including photocopying, recording or by any information storage and retrieval system without written permission from the author.

CONTENTS

Introduction 6

Large numbers 8

Chapter 1 – Cats 10

Chapter 2 – Too many sparrows – a failed strategy 14

Chapter 3 – Schieffelin, Shakespeare and starlings 17

Chapter 4 – The silent forest and parachuting mice 21

Chapter 5 – Minamata disease and "cat dancing disease" 25

Chapter 6 – Black summer in Australia 32

Chapter 7 – Japanese Knotweed 37

Chapter 8 – Bhopal gas leak 42

Chapter 9 – Bowl of oil and a feather 47

Chapter 10 – The Kudzu plant 52

Chapter 11 – Deepwater Horizon disaster 57

Chapter 12 – Ohio train derailment 63

Chapter 13 – Great emu wars - the army lost 68

Chapter 14 – Exxon Valdez oil disaster 73

Chapter 15 – One of the planets worst environmental disasters The Aral Sea 77

Chapter 16 – America's largest environmental disaster The Dust Bowl 83

Chapter 17 – London smog (smoke + fog = smog) 93

Websites 98

Arnold Lawson, a retired teacher, was the head of biology and the head of the science faculty in a large comprehensive school in Sheffield. From an early age and living on the edge of the English Lake District with its outstanding landscapes and its diversity of flora and fauna he has always had a great awareness of the environment. These interests have taken him from the forests of New Zealand to the Amazonian jungle of Brazil. His other interests include gardening, fell walking and landscape photography.

for
Madeline, Kathryn and Hazel

Introduction

Disasters have always occurred on earth. Some of these are natural disasters which are not connected to human activities, e.g. volcanoes and earthquakes. The worst natural disaster took place about 66 million years ago when an asteroid hit the earth and exterminated about 70% of all species.

In 1556, in Shaanxi, China, the deadliest earthquake ever recorded killed 830,000 people and such was the force of the earthquake it flattened mountains, moved rivers, caused massive floods and started fires that lasted for weeks.

In recent times, in 2010, an earthquake in Port-au-Prince, Haiti, killed 316,000 people, destroyed 250,00 houses and demolished 30,000 shops and business premises.

On Boxing Day, 26th December 2004, there was a massive earthquake off the coast of Sumatra, Indonesia. It was recorded as the third most powerful earthquake ever recorded since serious records were started in 1900. A record 30 metre tsunami wave devastated the coasts of 14 different countries causing the deaths of 228,000 people. The earthquake was so severe that it caused the earth to shake one centimetre!

The examples described above are only three of 5,700 major earthquakes recorded over a period of 4,000 years. In 2024, 1,374 major earthquakes were recorded in the world. Every day there are more than thousand earthquakes but many of these are too small for us to feel but can be recorded using scientific equipment.

Obviously, earthquakes can have catastrophic effects on the environment, but they are not caused by human activities.

Disasters caused by human activities are known as environmental disasters. There have been thousands of environmental disasters that have resulted in changes to the flora and fauna of ecosystems and have had an effect on human populations. There are different types of environmental disaster. Some involve air pollution; or it may be oil spills from ships or oil wells; it may be industrial when toxic chemicals escape from factories into the environment: sometimes environmental disasters occur when animals or plants are introduced to a new environment; poor agricultural practices can have disastrous consequences of the environment; nuclear accidents involving radiation leaks can create very dangerous situations.

In this book you will read about some of the world's worst environmental disasters and how they have affected ecosystems throughout the planet.

An ecosystem is all the plants, animals, fungi and bacteria living in a specific area. All these organisms are closely interconnected through food chains and food webs. Natural and environmental disasters can interfere with the delicate balance of ecosystems, and this can have disastrous consequences.

Large numbers

In this book you will come across some very large numbers. The information given below may help you understand these very large numbers.

Million – 1,000,000

How big is a million? Some people are millionaires, but we have very little idea of what this really means.

> If you were to count to a **million** at one number per second with no breaks it would take you **11 days, 13 hours, 46 minutes, 40 seconds**

Billion – 1,000,000,000 (a thousand million)

> If you were count to a billion at one number per second without a break it would take you 31 years, 251 days, 7 hours, 46 minutes, 40 seconds

**Area – this is measured in hectares (ha) There are 100 hectares in a km². ** The average size of a Premier Division football pitch is 0.6 hectares. You can fit 480 football pitches into one square mile. You can fit 100 football pitches into one square kilometre.

In the Australian Black Summer of 2019-2020, bushfires covered 24 million hectares of land. This is equivalent to 40 million football pitches!! This area of bushfires is greater than the size of England, which has an area of 13 million hectares (21 million football pitches).

Volume – this is measured in litres or cubic metres.
There are 1000 litres in 1 cubic metre or 1m^3.

An Olympic sized swimming pool is 50m long, 25m wide and has 10 lanes. This pool holds 2,500,000 litres of water or 2,500m^3. In the Mariana dam disaster in Brazil in 2025, the volume of mining waste released was 43.7x10^6 cubic metres or the equivalent of 17,480 Olympic swimming pools.

Length – this is measured in kilometres or km (there are 1.6 kilometres to a mile).

A healthy person takes between 10 and 15 minutes to walk a kilometre, but it depends upon the age, fitness, stride length of the person and of course upon the terrain of the walk. In the Mariana dam disaster in Brazil in 2025, the pollutants spread 668 km down stream. To walk this distance nonstop without any breaks and in a straight line would take you 167 hours or almost 7 days!

Chapter 1

Cats

You either like them or hate them! Cats have been domesticated for thousands of years and there is no doubt cats provide companionship for millions of people.

However, given the opportunity they are very nasty killing machines. With sharp teeth, sharp claws, patience, stealth, and with excellent night vision they make formidable predators. Much research has been carried out in the United Kingdom, America, Australia and other countries to investigate the impact that cats have on wildlife and the environment. The results show that your friendly pussy cat is not as innocent as you might think.

Biologists have discovered that domestic cats are responsible for killing more than 2,000 different species of prey – more than any other predator on earth! This included 981 different species of bird, 463 species of reptile, 431 species of mammal, 119 species of insect and 57 species of amphibian. In 2023, biologists discovered that 347 species of birds killed by cats were endangered. The Hawaiian Crow, the New Zealand Quail and the White Footed Rabbit Rat are now extinct in New Zealand because of cat predation.

Felmersham is a small village in the county of Bedfordshire, England. In the 1980's biologists collected information over a period of one year about the 70 cats living in the village and which animals they caught. Between them the cats brought home 1090 prey items, including 535 mammals, 297 birds and 258 unidentified animals. 30% of all sparrow deaths in the village were caused by cats.

Between 1st April and 31st August 1997, a survey of the number of animals brought home by cats in Great Britain was carried out by the Mammal Society. The results of the survey were devastating. 14,370 prey items were brought home by 986 cats living in 618 households. Mammals accounted for 69% of the prey, birds 24%, amphibians 4%, and reptiles, fish and invertebrates about 1% each. They also found that cats with a bell around their neck brought home fewer mammals and households with bird feeders encouraged cats to bring home more birds. Using the information from the survey biologists calculated that the **British cat population of 9 million, brought home 92 million prey items, including 57 million mammals, 27 million birds, and 5 million reptiles and amphibians in the period of the investigation.** The cat is the major predator of wild animals in Britain.

It is thought that the entire population of the flightless Stephen Island wren (an island off the coast of Queensland, Australia) was eliminated by a single cat – known as Tibbles and owned by the island's lighthouse keeper. The last bird was killed in about 1894.

In Australia the predation of animals by cats and foxes is staggering. Research has shown that **in Australia cats and feral cats kill one million birds a day**. (a feral cat is a stray cat, an escaped domestic cat, that is not owned by anyone). **Cats, both domestic and feral kill 1.4 billion animals a year including 700 million reptiles and 510 million birds**. The cats prey upon 338 Australian native bird species which is half of the number of bird species in the country.

Given the opportunity your friendly pussy cat is an extremely nasty killing machine.

Cats kill mice

Cats kill rabbits

Environmental Disasters

Drawing of the Stephen Island wren. It is thought that the lighthouse keeper's cat called Tibbles was responsible in 1894 for the final predation of the bird and its extinction.

Drawing of the New Zealand quail. Once common throughout New Zealand it is thought that feral cats were partly responsible for this bird becoming extinct.

Chapter 2
Too many sparrows – a failed strategy

You would not think that sparrows would be serious pest, but in China in the 1950's, this delightful little bird was responsible for the destruction of food crops on a massive scale.

In the 1950's Chairman Mao of China introduced new ideas in agriculture which he believed would lead to a prosperous future for the country. He decided that the Chinese people should destroy four pests – sparrows, flies, rats and mosquitoes. There was a problem – how do you kill millions of sparrows? Poisons could not be used – people could be killed; shooting the sparrows would use too much ammunition; bird traps could be used, but it would require too many.

The pest that Mao decreed should be destroyed was the tree sparrow (also found in the United Kingdom). Chinese scientists calculated that each tree sparrow ate 4.5kg of grain a year and for every million birds killed there would be enough food for 60,000 people.

Not all the tree sparrows were killed and a million sparrows were imported from USA and from Russia and the tree sparrow population has now increased to hundreds of millions. When Chairman Mao was given this information, he decreed that every person in the country should be involved in killing tree sparrows.

On December 13, 1958, in the city of Shanghai people shouted, waved flags and made as much noise as possible. The birds flew into the air, but they were not allowed to settle and they fell out of the sky exhausted and were killed. People destroyed tree sparrow's nests and their eggs smashed. These methods of eradicating tree sparrows continued for months.

Guangdong Province, China, killed 31 million sparrows in a year
Anhul Province, China, killed 135 million sparrows in a year
ShanghaI, China, killed 600,000 sparrows in a day
China killed 2 billion sparrows in a year

After one year of sparrow killing farmers, the Chinese Government and Chinese citizens breathed a huge supply of relief. Crop production increased and more food was available for everyone. Unfortunately, this was not to last. Chairman Mao failed to realise that tree sparrows also ate insect pests and in particular locusts which feed on agricultural crops. With no predators the pests and locust population increased so much that crop production and famine raged in China. It is estimated that 10-30 million Chinese citizens died in the Great Chinese Famine between 1959 – 1961. In April 1960, Chairman Mao stopped the Great Sparrow Campaign and sparrows were replaced with bed bugs on the pest list.

Tree Sparrow – Chinese people killed 2 billion of these birds in the 1950's to protect their food crops. Unfortunately, this strategy did not solve the problem and famine followed.

Chinese people stopping the tree sparrows from landing. The birds became exhausted fell to the ground and were then killed.

A cart carrying the bodies of tree sparrows killed in a failed effort to increase food production.

Chapter 3
Schieffelin, Shakespeare and starlings

Eugene Schieffelin (1827 – 1906) was an American ornithologist. In 1877, at the age of 50, he became the chairman of the American Acclimatization Society whose main aim was to introduce non-native animals and plants into America for economic and cultural reasons. He was fanatically obsessed with William Shakespeare, the English dramatist. Schieffelin's main aim was the introduction into America all the birds that were mentioned in the writings of Shakespeare.

Attempts by Schieffelin to introduce bullfinches, sparrows, skylarks and nightingales into America had failed, mainly because it was too cold for them to survive. Schieffelin would have had problems in introducing all Shakespear's birds into America – in his writings he mentions over 60 different species of bird!

On March 6th 1890, Schieffelin and his servants released 100 starlings (imported from England at great expense) in Central Park, New York. In 1891 he released a further 40 starlings. A few years later only 32 of the original 140 had survived - but that was enough, unknown to Schieffelin the damage had already been done. Starlings eat a wide variety of plant and animal species, including centipedes, butterflies, grasshoppers; they bully other birds and take over their habitats; they are strong fliers and can easily find new feeding areas and they can produce many offspring. In over 130 years the original 32 surviving starlings reached a population of 200 million birds in 1970. However, by 2017 the population had declined to 140 million birds.

You may think that this is a success story, but this is not the case. Starlings in America cause $800 million dollars (£606 million pounds) of damage every year. They feed on crops, cherries, strawberries and on cattle feed. They gather in very large flocks known as murmurations which may contain thousands of individuals. When the flocks reach the ground, they decimate crops and cover the land with bird droppings which can contain many diseases, one of which can cause blindness in cattle, horses and dogs.

In 1960 an aircraft taking off from Boston airport in America flew into a murmuration of starlings, crashed and 62 passengers were killed – only 10 survived. Such is the damage caused by starlings that many attempts have been made to eradicate them – shooting, poisoning and trapping. In 1960, nine million starlings were poisoned in California to protect cattle feed. All methods have failed, and starlings continue to survive in all parts of America.

A satirical drawing of Eugene Schieffelin with his starlings in Central Park, New York

The starling introduced into America from England by Eugene Schieffelin on March 6th, 1890.

A murmuration of starlings can contain thousands of birds.

A murmuration of starlings about to land on a field. They can now destroy crops and leave large amounts of droppings on the crops and soil. Starling damage in America is estimated at $800 million (£606 million) every year.

Chapter 4

The silent forest and parachuting mice

The brown tree snake was accidentally introduced onto the island of Guam (an island in the Pacific Ocean) by being a stowaway on an American military cargo ship in the late 1940's or early 1950's. The snake is slender about four metres in length with a great ability to climb trees. It has a strong grip around branches, good balance and very quick reactions. Although it is venomous and its bite painful it is rarely fatal in humans. There are no natural predators of the snake on the island of Guam.

The forests of Guam provided the snake with lots of food – birds eggs and bird chicks. So successful is the snake in preying upon birds that in the 1980's ten of the twelve native species of bird had been eradicated. The forests no longer have bird song – "Guam is eerily silent". With so much food available and no natural predators the snake population has reached enormous numbers. There are 5,000 snakes per square km. This is one of the highest concentrations of snakes ever recorded anywhere in the world. Thousands of the snake are even found at Guam airport and seaport.

However, the brown tree snake has created more problems. It has the ability to climb electricity poles and along electricity cables. In the last seven years it has caused $4.5 million (£3.35 million) of damage to electricity systems and over 1,600 power failures.

In 2021, the USA government announced that $4.1 million would be made available to control the brown tree snake. Several methods have been introduced to control the snake, but the most bizarre involves parachuting dead mice containing paracetamol into the Guam forests. In 2021 the United States dropped by parachute 2,000 dead mice onto the forests of Guam. Each mouse contained 80mg of paracetamol, a common painkiller that is fatal to the snake. The dropping of the baited mice has continued and between 2018 and 2023, 46,000 mice baits have been dropped by helicopters. In 2023 the USA government announced a sum of $3.6 million (£2.6 million) for further eradication of the snake. Studies have shown that this method of eradication is beginning to be successful.

To complicate the problems of the brown tree snake even further, studies have shown that the regeneration of forests on the island have been reduced. Forest birds distribute tree seeds around the island – no birds, no regeneration.

The brown tree snake is a perfect example of how an invasive species can damage an ecosystem and destroy other species.

The brown tree snake has created great ecological damage to the island of Guam.

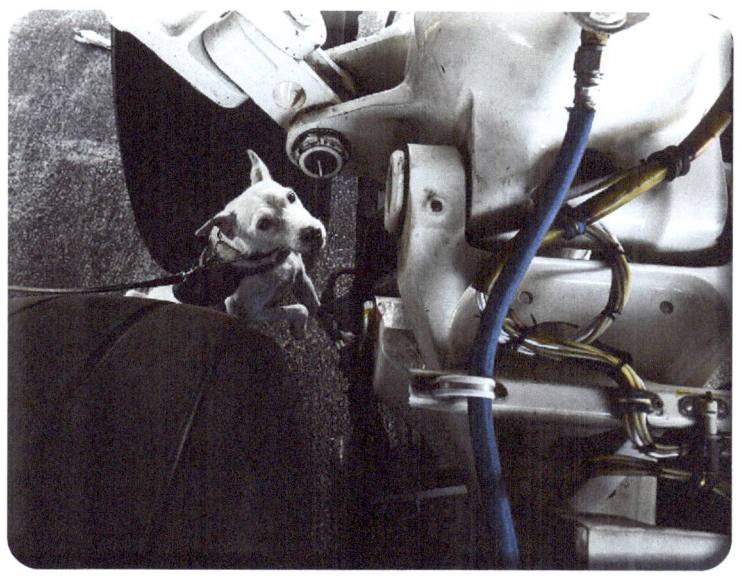

Specially trained dogs search for the snake at Guam airport.

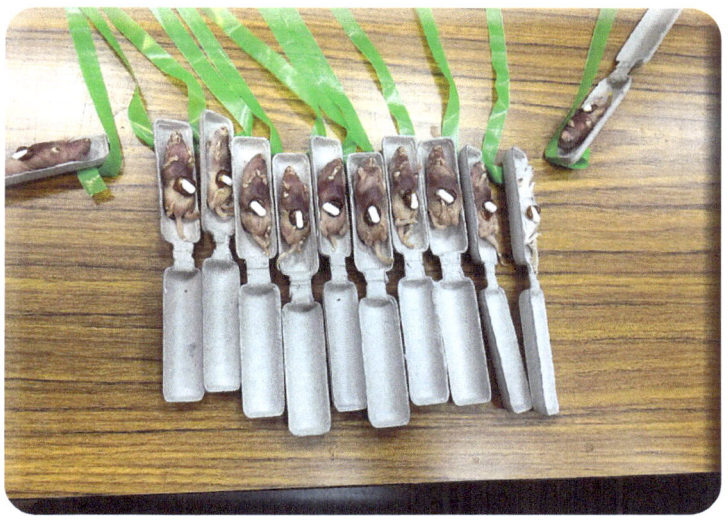

Dead mice containing paracetamol are dropped by parachute to kill the snake by poisoning

The brown tree snake has been responsible for the eradication of the Guam kingfisher (above) on the island of Guam. This bird is now only found in zoos. The Guam rail bird (below) has also been eradicated from the island, but efforts are being made to introduce it again in protected areas of the island.

Chapter 5
Minamata disease and "cat dancing disease"

Mercury is a chemical element and is the only metal that is a liquid at room temperature. It is however extremely poisonous especially mercury vapour. Mercury is listed as the third most toxic element in the world (after arsenic and lead). Even chemicals containing mercury are very poisonous. However, it is widely used – some thermometers contain mercury, it is used in some explosives, and some small batteries and electrical switches contain this element.

The first account of serious mercury poisoning was written in a newspaper called the Naval Chronicle, which was written in April 1810, over 200 years ago. On the 13th March 1810, 700 boxes of mercury metal, to be transported to South America, were loaded onto the ship H.M.S. Triumph. The liquid metal was contained in leather bags each holding 23 kg (50 lbs). The bags were packed into wooden barrels which were enclosed in wooden boxes and stored below deck. The total weight of mercury in the ship was 15,875 kg (17.5 tons) The conditions below deck were warm, wet and without ventilation. In these conditions the boxes and barrels rotted, and several tons of mercury escaped and flowed below deck. It even entered the ship's food containers and 3,600 kg (7,940 lbs) of biscuits were destroyed. Within three weeks of the chemical escaping cases of mercury poisoning appeared amongst the sailors.

A drawing of H.M.S. Triumph. The ship was launched in 1764 and was a 74 gun ship that was involved in successful battles. In 1771 Horatio Nelson, aged only 12 years was a sailor on board the ship. It was on this ship that severe mercury poisoning was reported.

The number of sailors suffering from mercury poisoning depended on where their living quarters were situated. Sailors who lived most of the time below deck suffered terribly from mercury poisoning. To begin with many suffered from ptyalism, a large increase in the amount of saliva in the mouth. This was followed by mouth ulcers, burning sensation in the stomach, vomiting, diarrhoea and paralysis. Some of the officers suffered the most. Their heads and tongues swelled so much that the eyes closed and the space between the cheeks and the collar bones was filled. Seamen lost their teeth and suffered from gangrene of the tongue and cheeks. Over 200 sailors suffered from mercury poisoning. H.M.S. Triumph also had on board sheep, pigs, goats, chickens, cats and a dog, a canary, rats, mice, and cockroaches - all died. All the metalwork on the ship, copper bolts, brass fittings, ironwork, even gold watches became covered in a coating of mercury.

On 22nd April the sick were transferred to hospital ships. H.M.S. Triumph was sent to Gibraltar and when it arrived on 6th May the ship was cleaned of mercury.

Mercury poisoning is now called Minamata disease, named after the city of Minamata in Japan where the release of large amounts of a chemical containing mercury was released from a chemical factory in 1956 and affected thousands of people.

The chemical factory in the city of Minamata, Japan, where chemical waste from the factory was deposited into a river which flowed into Minamata Bay. Here the chemical entered the food chain and caused severe methyl mercury poisoning in the local populations and cats.

In 1908 a chemical factory was opened in Minamata, Japan, to manufacture fertilizer. Gradually over the next 50 years the factory was developed to make different chemicals so that in the 1940's and 1950's it became the most advanced factory of its kind in Japan. Waste chemicals from the factory were released into the local river which flowed into Minamata Bay. One of the chemicals released in 1951 was called methyl mercury.

Listed below is a diary of events associated with methylmercury poisoning In Minamata.

1950 – Strange behaviour in cats reported. Cats had convulsions and appeared to go mad which was known as "cat dancing disease". Crows fell out of the sky, seaweed did not grow on the seashore and many fish were found dead.

21 April 1956 – First patient, aged five years, admitted to hospital with difficulty in walking, speaking and with convulsions.

23 April 1956 – Two girl sisters admitted to hospital with same symptoms as the first patient.

1 May 1956 – Officially an unknown disease was identified and labelled as Minamata disease.

24 August 1956 – Scientists from a university found other symptoms in patients. There was a loss of sensation in hands and feet, unable to grasp objects or fasten buttons. The voice changed, patients could not see clearly, and their hearing was poor. Patients could not swallow properly and had convulsions.

October 1956 – 40 people were found with the above symptoms and 14 died. This was a staggering fatality rate of 35%.

November 1956 – Discovered that people suffering from the disease all lived in villages on the shore of Minamata bay. They ate large quantities of shellfish (prawns, shrimps, lobsters, crabs and marine snails) and fish. Cats also ate these animals given to them as scraps of food. The chemical factory was suspected as being the source of the chemicals involved in the disease.

February 1959 – Large quantities of mercury compounds found in the shellfish and fish eaten by the villagers. Research scientists now had the evidence that Minamata disease was caused by food poisoning with methyl mercury entering the body through eating contaminated shellfish and fish. On the 21st October 1959 the factory was ordered by the Japanese government to install better waste water treatments and therefore reduce the methyl mercury outflow from the factory. On the 19th December 1959 the president of the factory drank the water supposedly from the new water treatment system to prove that the water was safe to drink! However, the water was not from the new system and in the Minamata disease trial researchers found that the new water treatment system was totally useless!

1960's – In 1960 it was reported that there were no more new cases of Minamata disease and this was because the eating of shellfish and fish from Minamata bay was banned and because new treatment facilities of the pollutant were now in place in the factory. However, this was not true. Waste water containing the methyl mercury continued to be released and people were still being poisoned and killed by the pollution. Researchers discovered extremely high levels of methyl mercury poisoning in most people in the Minamata Bay area. However, the authorities did not release the results of the research.

26 September 1968 – Finally after twelve years after the discovery of the disease, the Japanese government officially declared that Minamata disease was caused by methyl mercury poisoning from eating contaminated food containing the chemical. This brought great relief to the victims and their families and identified the polluter as the chemical factory which for years had denied that they were to blame. Now the families concerned could concentrate on the amount of compensation they could expect.

April 1969 – A committee was formed to calculate the compensation to be given to the victims and their families. This money would be paid by the factory.

March 1970 – The committee decided that the following payments should be paid to those people who had suffered from the disease (all the payments have been changed into £'s sterling and 2023 equivalent.

        Dead patient - £38,000

        Surviving patient - £2,600 to £3,800 per year.

However, the committee came in for much criticism that the payments were not big enough! Therefore, a final payment scheme was finalised on the 27th May 1970.

Dead patient - £32,000 to £75,500.

Surviving patients - A one off payment of £18,700 to £79,300, and an annual payment of £3,150 to £7,500.

However, the surviving patients were not yet finished and they took the company to court! The trial took nearly four years. IN 1973 the company was found guilty negligence and putting profits before health and safety. The company was ordered to pay:

Dead patient - £450,300

Surviving patient - ££402,700 to £450,300

The total compensation was £23, 212, 963- the largest sum of money ever awarded by a Japanese court!

It was estimated in 2001 that 2,265 victims died and over 10,000 received compensation.

Even today victims of Minamata disease are still suing and receiving compensation for the disastrous release of methyl mercury that took place over 70 years ago.

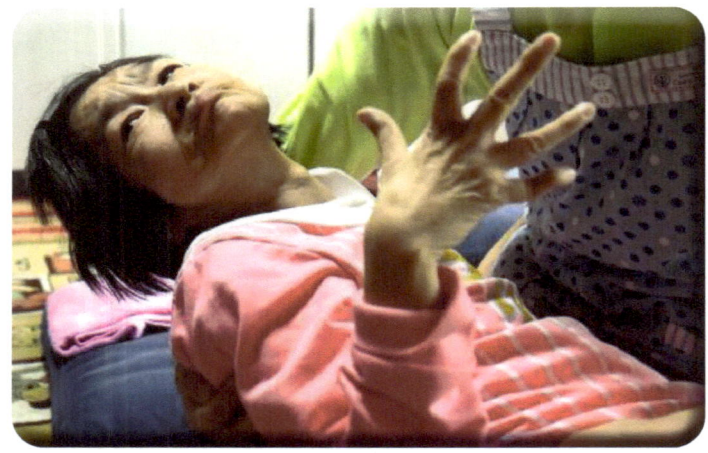

Photographs of people suffering from Minamata disease.

Environmental Disasters

There is no cure for people suffering from Minamata disease

Chapter 6
Black summer in Australia

Bush fires have always occurred in Australia – it is probably more prone to fires than any other country. They take place every year and although they destroy wildlife, the ecosystems always recover with the high temperatures being essential for the seeds of many plants to germinate. However, between 2019 and 2020 the bush fires were catastrophic causing incredible damage to ecosystems and all aspects of life in the affected areas of Australia.

Fire destroyed 24.3 million hectares of land. This is equivalent to the size of 34 million football pitches. The area burned was greater than the area of the Netherlands, Denmark. and Switzerland combined.

3,500 buildings destroyed, including 2,779 houses.

Fires killed 34 people. Smoke killed 400 people and 4,000 people in hospital suffering from smoke inhalation.

Smoke from the fires blown 11,000km (6,200 miles) across the Pacific Ocean. Smoke clouds appeared in the skies of Chile and Argentina.

Estimated that 700 million tonnes of carbon dioxide were produced in the fires.

Fire fighters from New Zealand, USA, Singapore and Canada brought in to help Australian fire fighters.

The Tasman glacier in New Zealand turned brown as soot from the fires was deposited in the ice.

On the 8th June 2020 it was estimated that one billion animals were killed. This included mammals, birds and reptiles). Trillions of invertebrates were killed.

On Kangaroo island (located off the south coast of Australia) there was severe damage. 52% of the island's land was destroyed and 80% of the islands 50,000 koala bears were killed.

The bush fires of 2019 to 2020 were the most expensive natural disaster ever to occur in Australia. The damage to buildings, farmlands, and tourism was to cost an economic loss of 78 to 88 billion Australian dollars (equivalent to £41 billion pounds).

Animals and plants have evolved over time to survive these fires, but in the Black Summer some of the fires occurred in areas which do not usually have fires. However, recovery of the ecosystems has taken place in some areas, albeit very slowly. Bush fires have been essential in shaping Australian ecosystems. Australia's animals and plants have developed many strategies to survive bush fires. However, the fires were so serious that many species have now become very rare or even extinct.

Bush fires in Australia occur every year, but in 2019-2020 the fires were so severe that it caused incredible damage in many parts of the country

An aerial photograph showing the extent of bush fires in Australia in 2019 to 2020.

Millions of animals were killed, and many were rescued like this koala with severe burns to his fur and skin.

Environmental Disasters

The Australian bush after the fires of 2019 to 2020.

Thousands of sheep were killed on Kangaroo Island, Australia during the Black Summer of 2019-2020.

25,000 koalas were killed.

The wombat is a one metre (three feet) long herbivorous marsupial. It digs long burrows up to 200 metres in length. It has a backward facing pouch so that when digging the soil does not enter the wombat's pouch where its young called a joey is protected. During bush fires many animals take refuge in the burrows and escape the fires.

Many plants in the Australian bush need fire to stimulate growth. The pink flannel flower shown above is an example of such a plant. A chemical found in the smoke of the bush fire is necessary for the seeds of the plant to germinate.

Chapter 7

Japanese Knotweed

It is not very often that a plant can influence whether a house is sold or not – but this is the case with a plant called the Japanese Knotweed. In the United Kingdom £41 million pounds are spent every year trying to eradicate the plant.

The plant was introduced into this country from Germany in 1850 and sold to plant collectors and plant nurseries as an ornamental and attractive plant. It has now spread throughout the country and is an extremely invasive plant causing severe damage to houses and other buildings and is described as one of the world's most invasive species.

The plant can grow four metres (thirteen feet) in 10 weeks and with an extensive root system that can reach a depth of six metres it is a very difficult to eradicate. In Japan and south east Asia the original home of the plant there are 186 species of insect and 40 species of fungi that feed on the plant and keep it under control. These organisms cannot be used in this country as they would become invasive species and could create even more problems.

Such is the seriousness of this plant that it is very difficult to obtain a mortgage and buy a house if this plant is growing in the garden. There are heavy fines and even jail for anyone selling a house and not informing the buyer that the plant is present in the garden. In 2022 a person was fined £32,000 for not informing the new householder that Japanese knotweed was growing on the property. If you cut down the plant and take it to the local tip you could be accused of spreading the plant, fines up to £5,000 with a possible two years in jail.

The plant forms a very thick canopy that does not allow other plants to grow. The extensive root system can grow through cracks in concrete, tarmac and brickwork and the cost of controlling this invasive species in the United Kingdom can cost up to £2,000 for an area 5mx4m. It was found growing on the sites for the building of the velodrome and aquatic centre in London for the 2012 Olympic Games and cost £70 million pounds to remove.

However as more knowledge is accumulated about Japanese Knotweed some of the laws dealing with the plant have been changed e.g. the distance between the plant and the property has been reduced to three metres.

Environmental Disasters

Japanese Knotweed, described as one of the world's most invasive species and can cause extensive and expensive damage to buildings. The plant can grow up to four metres (13 feet) in 10 weeks and can create real financial problems for the unsuspecting houseowner.

Japanese Knotweed has an extensive root system that can reach a depth of six metres (nineteen feet).

Japanese Knotweed splitting open a brick wall.

Japanese Knotweed breaking through the foundations of a building.

Japanese Knotweed growing through cracks in a concrete patio.

Chapter 8
Bhopal gas leak

The Bhopal gas leak is considered to be the world's worst industrial disaster. There is a constant battle between growing our food and the pests that feed on our crops. For farmers to produce enough food for the world's population, which in November 2024 was estimated to be 8,147,701,969 (just over eight billion) it is necessary to produce chemicals to control crop pests. These chemicals are called pesticides. The United Nations state that 40% of our food crops are lost to pests. This costs the world economy about $220 billion dollars (£163.8 billion pounds) every year.

On the 2nd December 1984 a chemical called methyl isocyanate was accidentally released into the air from a factory in Bhopal, India. This chemical is used to manufacture a pesticide known as sevin. In 1969 a factory was built in India to manufacture this pesticide using methyl isocyanate. Even before the accident in 1984 there were several complaints about poor safety and leaks in the factory. So, on the 2nd December 40,640 kg of methyl isocyanate gas escaped in a period of two hours. Strong winds blew this chemical over the city of Bhopal and 500,000 people in the city and surrounding villages were exposed to this highly toxic gas. The effects of this gas were terrible. People had bouts of coughing, severe eye irritation, burning sensation in the lungs with the kidneys and liver affected. It also affected newly born babies and pregnant women.

Doctors, nurses and health authorities were not prepared for such an outbreak of poisonous gas. It was not known how to treat methyl isocyanate poisoning. 170,000 people were treated in hospital with 2,258 deaths. Twenty-four years later people were still dying from the effects of the poisonous gas. Many people were not killed but suffered later in life from cancer and blindness. The incident also had other effects – 2,000 dead buffalos and goats were collected and buried; thousands of other animal were killed. Fishing was prohibited as the gas entered fishing areas.

You may think why such a dangerous chemical was made. The pesticide made from this chemical was though to be a major breakthrough in the control of insect pests. The pesticide which was first advertised in 1956 had many attractive properties – it controlled over 500 insect pests; it did not kill vertebrates; it did not persist in the environment; it did not accumulate in fats found in animals as happened with some pesticides; it is cheap to manufacture. However, it is toxic to humans when breathed in or swallowed and causes cancer. Although it kills mosquitoes it is highly toxic for bees which are essential for pollination in many food crops. It is still used in America but is banned in the United Kingdom and Europe.

WORLD'S MOST INDUSTRIAL DISASTER	
Where	Bhopal, India
When	2nd November 1984
Chemical released 40,640kg of a gas called methyl isocyanate	
Number of deaths in 2 weeks	8,000
Total number of deaths	15,000 to 20,000
Number of victims paid compensation	574, 366
Compensation paid to victims	150 million dollars (£114,330,000)
Poisoned more than 500,000 people	

Thousands of people died with methyl isocyanate poisoning. Hospitals and local health authorities could not cope with the large number of bodies killed by the gas.

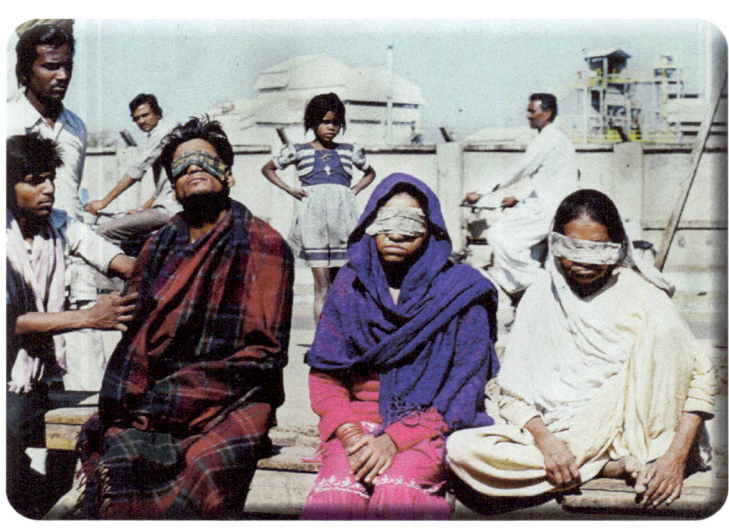

Severe damage to the eyes and blindness was one of the symptoms of methyl isocyanate poisoning.

Environmental Disasters

Thousands of cattle were poisoned and killed by the gas methyl isocyanate

Over 40 years later in January 2025 the Indian government stated that the site of the disaster had been cleared by removing 337 tonnes of toxic waste. The actual transport of the toxic waste was extremely complicated and required months of organisation and preparation. Trucks carrying the waste had to be leak proof and fire resistant. The trucks carrying 12 shipping containers had to transport the waste from Bhopal to Pithampur where the waste was to be incinerated over a period of three to nine months. The transport was so dangerous that the trucks were escorted by more than 50 vehicles which included fire engines, police vehicles and ambulances with 700 security staff. The 12 trucks were driven non-stop for seven hours and the roads were closed to all other traffic. The transport of the toxic waster cost $14.5 million dollars (£10.8 million pounds).

There were problems however as many people still believed that there was much toxic waste still in the ground of the factory and that chemicals were polluting drinking water.

Transporting 337 tonnes of toxic waste from the Bhopal disaster area 143 miles to Pithampur where the waste was to be incinerated. The waste is being removed over 40 years after the incident took place.

However, victim support groups were stating in 2024 that 150,000 people were still suffering from the inhalation of the poison gas. Chronic cancer, lung and breathing problems and nerve disorders were still present in the victims.

Chapter 9
Bowl of oil and a feather

Sometimes trying to control a pest using another animal (this is called biological control) can create very unusual problems. This was the case when rats in the 19th century caused much damage to sugar cane plantations in Jamaica.

Jamaica started growing sugar cane in the 16th century and in the 18th century was the largest sugar cane producer in the world. Although the Jamaican Rice rat was involved in the reduction of sugar cane production in the 19th century it was the accidental introduction of the Brown rat escaping from ships and living in sugar cane plantations that was to create catastrophic ecological problems for Jamaica. In 1872 William Espeut's Spring Garden sugar plantation was being devastated by the brown rat. On the advice of his wife who had lived in Ceylon (now called Sri Lanka) and had seen her pet mongoose kill rats she suggested that if Jamaica could import the mongoose, then this would reduce the destruction of her husband's sugar plantation. William Espeut wrote to the Jamaican government suggesting that the mongoose should be used to control the brown rats. Permission was granted and in February 1872, four male and four female Small Indian mongooses from India were released in his Spring Garden estate. Within six months it was reported that the Small Indian mongooses had done their job and was a huge success and there were fewer rats in the sugar plantations and more important there was an increase in plantation profits. In 1881 a law was passed protecting the mongoose. However, things were about to get much worse. The mongoose population had increased so much and with fewer rats for food, the mongoose now turned to other animals upon which to feed. They ate young pigs, lambs, kittens, puppies, poultry, birds which nested on the ground, small mammals, eggs, snakes and lizards. The mongoose was now regarded as the greatest pest ever to be introduced to Jamaica.

The situation was to get even worse. There was now a population explosion of ticks (a small species of spider) which previously had been kept under control by the animals that had been killed by the mongoose. All over the island people complained of being bitten by the ticks. In 1896, Henry Hesketh Bell, an administrator arrived in Jamaica. When he was shown his accommodation, he was surprised to see a small bowl of oil and a feather on a table. He was told that these were to be used to remove the ticks from his skin and clothing!

Not only was there an increase in the number of ticks there was also a plague of caterpillars – normally these would have been eaten by the lizards and birds, but they were reduced in number by the mongoose. In 1896 thousands of acres of vegetation and food crops were destroyed including fields of corn, potatoes and cassava. The caterpillars were probably from a species of day flying moths.

The reason for the tick population explosion was the killing of their predators by the mongoose. The mongoose still survives in Jamaica and the introduction of this species reminds us how easy it is for humans to alter the delicate balance of nature.

Environmental Disasters

The Small Indian mongoose, introduced into Jamaica with a devastating effect, to control the brown rat.

The brown rat escaped from trading ships and severely damaged sugar cane plantations and was responsible for the reduction and even extinction of other animal species.

The population of ticks which occurred in Jamaica and which oil and a feather was used to remove them from skin and clothing.

The caterpillars of this moth caused extensive damage to food crops in Jamaica after mongooses killed the predators of this insect.

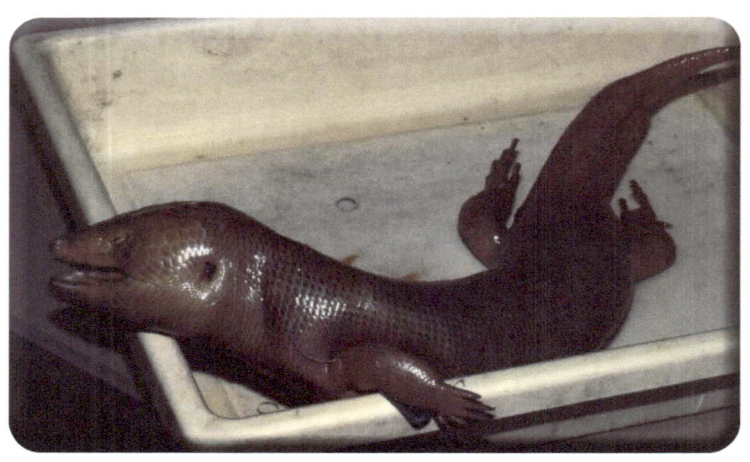

This 40cm to 50 cm long unusual looking lizard is the Giant Galliwasp. The mongoose was responsible for the extinction of this animal in the mid 1800's in Jamaica.

The eggs of the Jamaican petrel were eaten by the mongoose and the last sighting of this bird was recoreded in 1879

Chapter 10
The Kudzu plant

The Kudzu plant is America's most famous and damaging weed. It is in the top ten of the world's most invasive species. The plant was first introduced into America at the World Fair Exhibition in Philadelphia, America, in 1876. The plant, a native of China, Japan and Korea was marketed as an attractive plant for the garden and good forage for livestock.

In the prairies in America decades of intensive farming of wheat and high temperatures, low rainfall, severe drought, strong winds and poor agricultural management created an area known as the dust bowl. In the 1930's in America there was a series of dust storms that severely damaged the agriculture and the ecology of land known as the prairies. The soil was turned into "dust" creating "black blizzards" with almost zero visibility and could be blown for hundreds of kilometres. The damage to this wheat growing area of America caused farmers to be losing $25 million (equivalent to $550 million dollars at 2023 prices, or £550 million pounds) per day in 1936. Between 1930 – 1940 , 3.5 million people emigrated from the prairies to other parts of America because of the dust storms and because they could not make a living from farming. The American government decided that to reduce soil erosion on the prairies the kudzu plant should be grown. There were good reasons to grow this plant; it was easy to grow; it grew very quickly at 30cm (12 inches) in a day; had a good root system to bind the soil. The plant has a taproot (rather like a carrot) that can be over three metres in length and weigh 180kg or 397lbs)! The government funded the growing of 85 million kudzu seedlings and by 1946 1,200,000 hectare of land had been planted with kudzu. (This area is larger than the county of

Yorkshire, the largest county in England). There was even a Kudzu club of America, with a membership of 20,000 whose aim was to plant 8 million acres (3,237,000 hectares) of kudzu in the south of America. People became obsessed with the plant – a musical was written, and coffee houses, cafes and pubs were opened with a kudzu theme. However, by the 1950's farmers could not make any money from planting kudzu and the scheme was discontinued. The Kudzu plant grows so quickly that in the 1970's it was officially declared a weed, and at the present time the plant covers 3,000,000 hectares of land (this is almost the size of Belgium).

One of the most spectacular uses of the Kudzu plant was during World War 2. America had a United States Armed Forces base on the island of Fiji. The fast-growing Kudzu plant was planted around their equipment to give camouflage. The plant, now an invasive species in Fiji, creates the same problems as its growth in America.

Dust storms in the prairies of America in the 1930's. They were caused by drought, strong winds and poor agricultural practices which turned the soil into "dust". The kudzu plant was used to reduce soil damage.

The kudzu plant grows extremely quickly and smothers ground vegetation and even kills trees and destroys the ecology of the area.

Environmental Disasters

The Kudzu plant can easily grow over agricultural equipment and even houses

Kudzu roots are very large and in many parts of Asia their roots are used as food because they contain large quantities of starch. The starch is used to thicken gravy, in the making of pies and in stir fries. The roots have been used in Japanese and Chinese medicines for thousands of years.

In Japan, kudzu is thought of as a superfood.

Chapter 11
Deepwater Horizon disaster

Deepwater Horizon was a marine oil drilling rig with a computer-controlled system to keep the rig in the same position using propellors and thrusters. Using this system there was no need to anchor the oil rig to the seabed. This system is used on most marine oil rigs. The ten-year-old oil rig was drilling for oil and gas in the Gulf of Mexico located about 66 km (41 miles) off the coast of Louisiana. The seabed was 1600 metres (5,100 feet) beneath the rig and it had drilled downwards for 5,500 metres (18,000 feet) into rock to find oil and gas. The oil well was then sealed for future use using concrete. Unfortunately, the concrete seal was not strong enough and the pressure of the oil and gas in the well broke the seal and on April 10th 2010 the oil and gas exploded upwards into the oil rig causing a catastrophic fire.

What followed was to be the largest accidental marine oil spill in the world.

The explosion killed 11 people on the oil rig and injured 17. For 87 consecutive days oil and gas escaped from the oil well and caused very severe environmental damage. 210 million gallons of oil escaped from the well and this had disastrous effects on beaches, estuaries, wetlands, the fishing industries and tourism. The oil spill directly affected 180,000km2 (69,498 sq. miles) of sea and land. The oil well was capped, stopping the escape of oil and gases on 19th September 2010, but the damage had already been done.

During the oil leak many strategies were used to try and clean up the oil spill and protect the environment:

(1) Skimmer ships were used which scoop the oil out of the water and then separate the oil and water. The largest skimmer ship was called "the Whale" and sailed to the Gulf. At 340 metres long the owners of this giant ship stated that it could separate 300,000 to 500,000 gallons (1,140,000 – 1,900,000 litres) of an oil water mix a day. However, over two weeks it collected hardly any oil (because of poor sea conditions) and was rejected.

(2) Floating booms were used to stop the spread of the oil which could then be collected by small skimmer ships. 4,200,000 feet (795 miles or 1,300 km) of booms were used to contain the oil.

(3) One of the commonest methods used to stop the oil spreading in the Deepwater Horizon disaster was the use of dispersal chemicals (this is rather like using washing-up liquid for cleaning fat from plates). Unfortunately, the dispersants had a terrible effect on wildlife. In trying to clean up the pollution caused by the disaster 1,840,000 gallons (almost 6,965,000 litres) of these chemicals were used.

On the most demanding day of the disaster there were 47,949 people involved in controlling the pollution, 6,000 ships used, 82 helicopters and 20 aircraft used, and 2,063 skimmer ships involved. The effects of the oil spill and the use of dispersants were staggering on the flora and fauna of the area. Some of these effects are shown below :

800,00 birds died	167,000 sea turtles were killed
Between April 2010 and July 2014, 1,141 dolphins died.	32% of the population of laughing gulls in the area died.
20% of fish in the area had oozing sores.	50% of shrimps had no eyes or eye sockets.
8.3 billion oysters were killed.	25,900 marine mammals died.
27,000 to 65,000 of Kemp Ridley sea turtles were killed.	12% of brown pelicans in the area were killed.
The loss to the tourism industry was $27 billion (£20,115,000,000)	The loss to the fishing industry was $247 million (£184,114,000).

Thousands of turtle eggs were collected from turtle nests in the sand. They were taken to laboratories and incubated. When they hatched they were taken back to unpolluted areas. 14,000 loggerhead turtles were returned to the sea.

The Deepwater Horizon oil rig before the explosion. It has no anchors, but is kept in position automatically by using propellors and thrusters and satellite navigation.

The Deepwater Horizon oil rig immediately after the explosion on April 10, 2010. 210 million gallons (794,936,500 litres) of oil were released into the sea.

Environmental Disasters

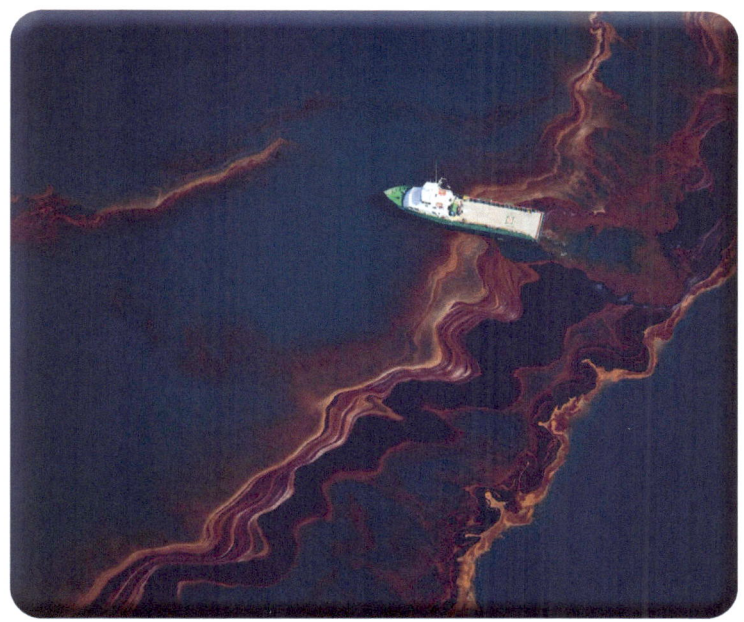

One of the 2,063 skimmer ships that were used to try and remove the oil from the surface of the sea However, because of very poor sea conditions they were mostly ineffective.

800,000 birds died from the disaster. The photograph shows a brown pelican covered in oil.

A turtle covered in oil rescued from the sea. Thousands of turtles were killed by the oil

Waste is safely secured and loaded for off-site disposal

Chapter 12
Ohio train derailment

Imagine a train 2.8km (1.76 miles) long with 151 waggons and weighing 18,000 tonnes with only three people on board! The goods train was travelling from Madison, Illinois to Conway, Pennsylvania, USA, when on 3rd February 2023 at 8.55pm, it crashed. 51 waggons were derailed and 49 of them ended up in a pile and caught fire. 12 of the derailed waggons contained in total 305,350 gallons (1,155,875 litres) of extremely hazardous chemicals. The chemicals released into the atmosphere from the accident included hydrochloric acid, phosgene and vinyl chloride. Phosgene is an extremely toxic chemical and was used as a chemical weapon in WW1 where it was responsible for 85,000 deaths. Vinyl chloride is an extremely dangerous cancer producing chemical used in the manufacture of PVC (polyvinyl chloride) and 115,000 gallons ((over 435 thousand litres) of this substance was released in the accident. PVC is used to make plastic pipes, electrical insulation, packaging and many other items. 36,300 million kg (40 million tons) of PVC are manufactured every year.

Because of the serious nature of the accident 17,500 people living within a 3km (2 mile) radius of the accident were immediately evacuated. A second evacuation of at least 25,000 people took place a day later.

Firemen used two million gallons (over 7.5 million litres) of water to extinguish the fire. However, the water run-off from the fire became contaminated with the chemicals and created many problems. The water reached streams and rivers and had devastating effects on the wildlife. Over 43,700 freshwater fish mostly small died. The soil around the crash became contaminated and 167,000 tons of soil were removed. 39 million gallons (over 14.5 million litres) of contaminated water was removed from the site. The contaminated soil and water were then transported to waste disposal facilities in the states of Ohio, Indiana, Texas and Michigan. There were many complaints about the transport of such contaminated materials from members of the public.

Investigations into the cause of the Ohio train derailment found that the fire was started because of an axle and wheel on one of the waggons overheated and set fire to one of the waggons which was carrying plastic pellets. The fire then spread to other waggons and explosions took place.

Although the fire was eventually controlled the effects on the populations of nearby villages and towns were serious. The chemicals produced during the fire contaminated homes weeks after the accident and seeped into curtains, chairs and covered the walls. Over 40 families had to move home and lived in hotels paid for by the firms involved in the accident. Many residents reported health problems including rashes, numbness, tingling in the mouth, pains in the ears and eyes. Blood and urine samples for some of the residents tested positive for vinyl chloride. An area of 540,000 square miles (this is an area larger than Peru) over 16 American states received contaminants from the disaster.

Experts predicted that it would take years to complete the clean-up. In 2024 the cost of the Ohio train derailment was $1.1 billion dollars (£845 million pounds) and insurance claims cost $101 million dollars (£78 million pounds).

Environmental Disasters

On 3rd February 2023, a freight train 2.8km (1.7 miles) in length derailed in East Palestine, Ohio, USA, caught fire and released deadly chemicals into the environment. The population of nearly 5,000 suffered from this environmental disaster.

Smoke containing many hazardous chemicals produced when the Ohio fright train derailed and then exploded.

Aerial photograph showing some of the waggons that were derailed.

The surrounding soil was contaminated and had to be removed

Environmental Disasters

Many homes in the area close to the accident had to be thoroughly cleaned to remove poisonous and carcinogenic chemicals.

Convoys of large trucks removed 167,000 tons of contaminated soil to waster dispersal sites. The convoys were accompanied by police, fire engines and ambulances.

Chapter 13

Great emu wars - the army lost

The emu is the second largest bird in the world (the largest is the ostrich) and is found in Australia. It reaches 1.5 m (5 ft) in height and weighs 45 kg (100 lbs).

The emu eats plants, grasshoppers, beetles, and other ground dwelling invertebrates. It also eats stones!

The emu has about 0.7 kg (1.6 lbs) of stones in its gizzard which is part of its digestive system. The stones help to grind the food into smaller particles to help digestion.

The emu is the only bird in the world that has calf muscles. This helps the bird to run at speeds upto 48 kmph (30 mph).

The emu has three toes which help to grip the ground when running when its stride is about 275 cm (9 feet). The middle toe is the largest and the most powerful. The toe and claw are 15 cm (6 inches) in length and can be used in defence if attacked.

Environmental Disasters

A group of emus is called a mob. The maximum size of a mob is about 20 birds. When good feeding grounds are found they gather in much larger groups but are usually spread out.

The emu was involved in rather bazaar and failed attempts to control their population in Australia in the 1930's.

At the end of World War 1 in 1918, about 5,000 Australian soldiers were given land in Western Australia by their government for farming. In 1929 there was an enormous financial crash in America (known as the Wall Street Crash) and this had a disastrous economic effect throughout the world, and Australia was no exception. Australia suffered (as did other countries) with high unemployment, poverty and low wages. In Western Australia farmers were encouraged to grow wheat to provide food for the communities. In 1932 there was a mass migration of 20,000 emus into the wheat growing areas and they ate and spoiled the wheat crops. The emus also pushed down fences and rabbits entered the fields and caused further problems.

The solution to this problem was easy and simple - shoot the emus. With the farmers being soldiers they were used to handling machine guns and were accurate marksmen. The emu war commenced in October 1932 when the army and farmers were issued with two machine guns and 10,000 rounds of ammunition. However, it rained on the first day of the slaughter and no birds were killed! On the 4th November the machine gun jammed and only 12 birds were killed! A few days later it was decided to mount the machine gun on a truck and follow the birds, but the land was so rough that the marksman was unable to fire the gun! By the 8th November, 2500 rounds of ammunition had been fired but it was reported that only 50 birds had been killed. When the birds heard the gun shots they scattered and it became more difficult to kill them. On the 13th November 40 emus were killed, but by 2nd December the success rate had improved and on the 10th December 986 birds had been slaughtered and 9,860 rounds of ammunition had been fired.

Killing emus was not easy. They have thick feathers and very tough skin. The military found out that an emu could easily take five bullets before it realised it had been shot and that it took ten bullets to kill the bird! The military were so unsuccessful in killing the emu that they became quite a laughing stock, objects of amusement and laughter which relieved some of depression amongst communities. The failure of the military to kill the emus was ridiculed in the press not only in Australia but throughout the world. Conservationists criticised the methods that were being used to reduce the emu population

Because of the failure of the military the government decided to may farmers for killing the emus. In 1934 farmers were paid for killing 60,000 emus. The killing of such large numbers had very little effect on the population of emus. In 1943 and 1948 farmers asked for the military to kill emus because of the devastating effect the birds were having on the farmers crops. The government declined, but in 1950, 500,000 rounds of ammunition were released for the purpose of reducing the emu population. Today agricultural crops are protected from the emus by high fences. Emus still exist in the country but the population now remains stable.

The large-scale introduction of wheat farming in Western Australia upset the local ecosystems and this created new feeding grounds for the birds and the emus became pests.

Farmers were paid by the Australian government or each emu killed.

Emus are kept out of agricultural land by high fencing.

Using the Lewis machine gun to kill emus in Australia in the 1930's

Environmental Disasters

Chapter 14
Exxon Valdez oil disaster

On 24th March 1984, at four minutes past midnight, an oil tanker owned by the Exxon Shipping Company ran aground on Bligh Reef, Prince William Sound, Alaska. The hull of the Exxon Valdez was ripped open and **eleven million gallons (41,800,000 litres) of crude oil was spilled** and had catastrophic consequences for the environment in the area. The ship was carrying 53.1 million gallons of crude oil - 20% of the ship's oil was released! ***The oil slick covered 2092 kilometres (1,300 miles) of coastline.*** The oil spill hade devastating effect on wildlife, as the table below shows.

250,000 sea birds	
3,000 sea otters	**ALL**
250 bald eagles	**KILLED**
302 harbour seals	
22 orca whales	
Unknown number of salmon and herring	
Billions of salmon eggs	

The spilt oil also had a devastating effect on the fishing industry and on the fishing communities who depended on fish for their livelihood.

The Exxon Shipping Company paid out $4.3 billion dollars (equivalent to $10.4 billion dollars or £8.3 billion pounds in 2024) in clean-up costs, legal costs and fines. They also paid $1.8 billion dollars (equivalent to $4.4 billion dollars or £3.5 billion pounds in 2024) for habitat restoration.

More than 11,000 people worked to clean up the oil pollution in the sea and on the coasts. They skimmed oil from the water surface, sprayed chemicals to disperse the oil and washed the beaches with hot water as well as rescuing and cleaning animals covered in oil. Biologists discovered that using high pressure water hoses was effective in removing the oil spills, but this method of cleaning also damaged the flora and fauna of the area.

After such environmental disasters it is essential that biologists continue to survey the polluted areas to find out the long-term effects of the oil pollution. In this incident an area known as Mearns Rock was not cleaned. Every year since 1990 this rock has been examined, surveyed and photographed to see whether there have been any changes in the flora and fauna of this rock. They observed that after three to four years the rock was recovering from the oil pollution and flora and fauna were returning. Biologists also discovered that that in 1999 (15 years after the disaster) the population of sea otters and bald eagles had recovered to the same numbers as pre the disaster. However, in 2019 oil was still to be found on some of the beaches of Prince William Sound and that the population of killer whale populations had not recovered and that fishing for the Pacific herring remained closed.

In 1984, the Exxon Valdez oil tanker was the largest of the United States' oil tankers. At 300 metres (984 feet) in length, it could carry 1.48 million barrels (62,160,000 gallons) of crude oil.

Environmental Disasters

A dead whale washed up on the coast and of Alaska, which had suffered from the effects of oil pollution.

Contaminated waste collected from the effects of oil pollution from the Exxon Valdez oil disaster.

Mearns Rock

Every year since 1990 this rock has been surveyed, from when the rock suffered from the effects of oil pollution (upper photograph). The lower photograph was taken about four years later and shows that the growth of seaweeds had recovered.

Chapter 15
One of the planets worst environmental disasters The Aral Sea

In the 1960's the Soviet Union decided to develop some poor land into agricultural land , but to make this productive irrigation was required. Two rivers, the Amu Dary (one of the largest rivers in Central Asia) and the Syr Darya were diverted to provide water for the irrigation of the new agricultural development for the growing of cotton. The area to be irrigated was 7 million hectares (2656 sq. miles). These two rivers flowed into the Aral Sea, the fourth largest freshwater lake in the world with a surface area of 68,000 km2 (26,300 sq. miles). The surface area was larger than that of the island of Sri Lanka.

As water from these two rivers was diverted the water level in the lake began to fall. By 1992, the water level had dropped 15m (50 feet) and its surface area reduced by half. By 1999, the lake was only one quarter of its original size. In 2015, 90% of the Aral Sea had disappeared. As the water level of the Aral Sea fell more and more, islands appeared. One of the islands was of very serious concern was called Vozrozhdenya . The Soviet Union had used this island as a secret testing ground for biological warfare and tonnes of extremely dangerous toxic material was buried on the island. This material included bubonic plague, anthrax, smallpox, TB and tularemia (rabbit fever which can be life threatening in humans). The testing ground on the island was the largest biological warfare site in the world.

In 2002 the USA sent a team of workers and scientists to successfully clear the island of the toxic material. The clearing of the site took two months and cost five million dollars (four million pounds).

The reduction in the size of the Aral Sea brought about a dramatic environmental disaster:

1. With the loss of water in the lake , the concentration of mineral salts , fertilizers and pesticides increased. Consequently, the fish population of sturgeon, carp, barbel and roach declined.
2. Villages that relied on the fishing industry for their existence ceased to exist. Over 100,000 people became unemployed.
3. The mud at the bottom at the bottom of the which was once covered in water dried out and turned to dust. Winds blew this dust over large areas and because it contained salts, fertilizers and pesticides many health problems were created. There was an increase in asthma, throat cancer and kidney disease. Infant deaths increased and became one of the highest levels in the world.
4. The Aral Sea was an important stopping point for millions of migrating birds such as flamingos, pelicans and many other species. They all disappeared, and the birds had to find new feeding grounds.
5. The dust blew onto snow and ice, absorbed more sunlight and this encouraged quicker melting.

Realising the damage to the Aral Sea, the Kok-Aral dam was built, financed by the World Bank, to divert some river water into the lake. This has had some limited success and there has been an increase in water level. There has been reduced salinity of the water and an increase in fish populations. This has brought some economic relief to the area with an increase in the fishing industry, job creation, and a return of people to the deserted villages. There has also been an increase in the planting of grasses and shrubs to stabilize the soil and reduce dust storms.

This disaster was caused by very poor environmental planning and not understanding the complexities of ecosystems and that simply diverting water for agricultural development can create disastrous consequences for local economies, industry and people.

Aral Sea in 1985. It is 418 km (260 miles)long, 281km (175 miles) wide, average depth of 16m (33 feet), maximum depth of 67 m (220feet). The surface area of the lake is 67,000 km2(26,000 sq. miles)

Reduction in the size of the Aral Sea.

Fishing boat on the dried-up Aral Sea.

Fishing boat on the dried-up Aral Sea.

The Aral Sea has become a desert. It is a mixture of sand, fertilizers, pesticides and salt.

Dust storm in the Aral Sea area. The dust was responsible for serious health issues for the people living near the lake. There was an increase in the cases of asthma, throat cancers and kidney diseases as well as an crease in infant deaths.

In 1996, the country of Tajikistan issued stamps to highlight the plight of the disappearance of the Aral Sea.

Chapter 16

America's largest environmental disaster The Dust bowl

If you are interested in or fascinated by large numbers, then this is the chapter for you.

1. In 1862, 1901, 1909 the American government gave 160 acres of land to each settler. Most of these were inexperienced farmers.	**2.** In 1920's and 1930's there was a great demand for wheat in Europe because of WW1. This encouraged farmers to plough more land for wheat production. In the 1920's the weather was good for the growing of the wheat crop.

Dust Bowl

3. The grassland plains of America were now over-ploughed. The development of tractors and agricultural equipment helped considerably to over-plough the land. Deep ploughing occurred and this destroyed the fragile ecosystem of the grasslands. The grasses were essential in binding the soil structure and this was now lost.	**4.** In the 1930's there was a severe drought in the wheat growing areas. Temperatures increased, there was no rain and strong winds occurred. In 1929 the Great Depression occurred in America. There was mass unemployment, banks failed and savings were lost. Farmers lost money, this affected agriculture and farmers abandoned their farms. Losses reached $25 million per day in 1936 ($550 million per day at 2023 prices, or £409.6 pounds).

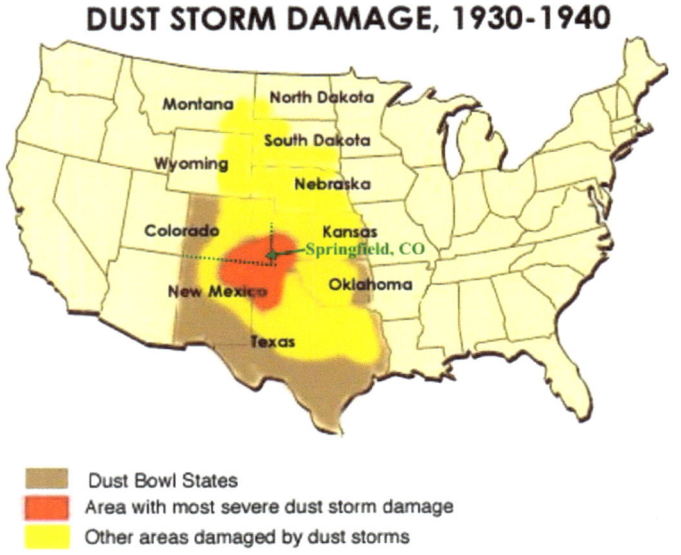

Springfield County was one of the hardest hit areas in the country

1933

In this year there were 39 dust storms. The storms were of different colours. Those from the state of Kansas were black in colour, those from Oklahoma were red and those from Colorado and Mexico grey, reflecting the colour of the soil. Heavy wet sheets were hung in front of doorways and windows to try and stop blown soil entering their houses.

9th May 1934

A dust storm 1500 miles (2,414 km) long, 900 miles (145 km) across and 2 miles (3.2 high) occurred. When it reached New York it was 1800 miles (2899 km) across and it was estimated to weigh 359 million tons. The sky became black and it was impossible to see the Statue of Liberty.

1935

850 million tons of topsoil was blown away from the southern states. 480 tons of topsoil were lost per acre of land. In 1935, 2.5 million people left their farms.

14th April 1935

A news reporter named Robert Geiger invented the name "Dust Bowl". On this day a dust storm over 200 miles (322 km) across and 2,000 feet (0.6 km) high travelled at 65 mph (104 kmph) from Oklahoma to Texas. As the storm moved south it became larger and eventually it was 1000 miles across. It was one of the worst storms in Dust Bowl history and was named the "Black Sunday Storm". The temperature at ground level fell by 100C. Vehicles could not move; visibility was almost zero and the storm lasted for 4 hours. 300,000 tons of soil were blown away. Twice as much soil was blown away as was dug out to make the Panama Canal and that took seven years to dig!

1935

The American Red Cross set up six emergency hospital in the Dust Bowl area. 17,000 face masks were distributed. Doctors and nurses visited hundreds of homes where people suffered from the effects of dust storms. In the state of Kansas, a dust storm forced cars to stop moving for eight hours. A basketball match was cancelled because players could not see the end of the court.

> **1936**
> In July 1936 a temperature of 500C was recorded in the state of North Dakota.

> In 1937 there were 134 dust storms.

> **Spring 1938**
> Rainfall started again at this time and by 1941 most areas which had suffered from the dust storms were receiving their normal amount of rain.

> **2022 – 2023**
> There was a reduction in rainfall in the 2022 – 2023 wheat growing area of America (the dust bowl area). As a result there was a reduction of 37% in the amount of wheat harvested. Some scientists have suggested that with climate change there could be possibly a return to the conditions of the dust bowl of the 1930's.

Environmental Disasters

Dust storm, April 18th 1935, Texas

Large dust storm, Colorado. May 1936.

"Black roller" dust storm, Oklahoma 1935

Environmental Disasters

Soil blown against building and destroying the home for habitation.

Farmers fleeing the Dust Bowl storms

Although the dust bowl created horrendous problems on the lives of adults and children there were many other effects. There was a population explosion of jackrabbits (also called "black-tailed jackrabbit", but not a rabbit at all, but a hare). Towns in the area organised "rabbit drives" where the animals were forced into enclosures and beaten to death with base-ball bats and clubs. In western Kansas 98,000 men, women and children were involved in 269 drives. An average of 923 rabbits were killed per drive (over 248,000 rabbits were killed in total). Over the area of the "dust bowl" it has been estimated that two million jackrabbits were killed. Some of these were now eaten, others were packed on refrigerated waggons and transported to other areas as a source of food. Millions of grasshoppers descended on the land and fed on any vegetation they could find. It was calculated that there were 23,000 grasshoppers per acre of land, (an acre is 4840 yd^2, this is approximately 70 yards by 70 yards).

Jackrabbits driven into enclosures and clubbed to death.

7,000 people died during the "dust bowl". Some died from starvation, but most died from breathing problems brought on by inhaling the dust, and asthma, bronchitis and influenza were common and the term "brown plague" was used to describe the symptoms. To reduce the effects in schoolchildren, they were often told to remain in school overnight to escape from the dust storms.

A teaspoonful of soil contains about 10,000 different species and they are essential for the recycling of nutrients, decomposition of dead species and improvement of soil structure – with the "dust bowl" all this microbial structure disappeared. The soil dust settled on plant leaves, reduced photosynthesis and transpiration and plant growth was affected. The dust covering on the plants was also responsible for the leaves absorbing more heat and caused wilting in plants.

In 1935, President Roosevelt's government decided that action had to be taken to stabilize the area of the dust bowl. The "Roosevelt Shelterbelt Project" was to plant 220 million trees in a 100 mile (160 km) wide and 1,000 (1609 km) mile long belt to halt the erosion of soil. This was to be carried out by boys, farmers and conservation workers.

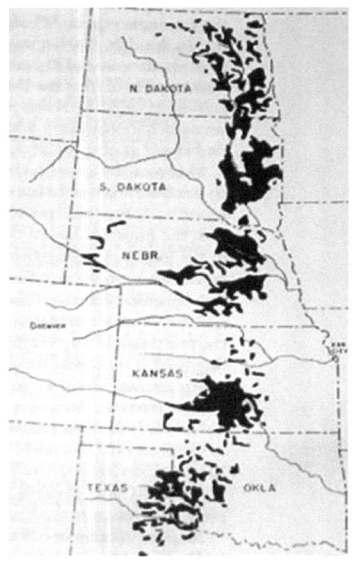

Map showing the areas of the "dust bowl" where trees were planted to reduce soil erosion.

Planting trees as part of the Roosevelt Shelterbelt project to stop soil erosion. The first tree, one of 220 million, was an Austrian Pine, and was planted on March 18th 1935 in Oklahoma.

Between 1934 and 1936 research has shown that 1.2 billion tons of soil was lost across 100 million acres (150,000 sq. miles, ie 395 miles (635 km) by 395 miles (635 km)).

A line of trucks 96 miles (154 km) long hauling 10 full loads a day would take a year to transport the soil from one side of the state of Kansas to the other – a total of 46 million truckloads!!

'The "Dust Bowl" was the most damaging and prolonged environmental disaster in American history'

Chapter 17
London smog (smoke + fog = smog)

Between 5th December 1952 and 9th December 1952, London was to suffer from a lethal environmental disaster. London had always suffered from fog, but the one in 1952 was probably the worst.

In 1273 the Parliament of Henry III passed a law prohibiting the burning of soft coal (poor quality coal which produced much smoke and nasty smells) because his wife, Queen Eleanor, had to flee Nottingham castle because of the fumes from the burning of coal. In 1306, the King, Edward 1, and his government passed what is thought to be the first environmental law when they banned the burning of coal if parliament was in session.

London was an expanding city, industry was thriving and smoke pollution increased. People had more money and could afford to heat their homes in cold weather and this further increased air pollution. Strangely people did not panic with this increase in pollution – London always suffered from pollution. However, in December 1952, London suffered its worst levels of pollution, and its inhabitants suffered four days of atrocious living conditions.

There were several factors which contributed to this disaster:

1. Unusually cold weather
2. Very heavy smoke pollution because households and the six London power stations burnt more poor grade soft coal (known as "nutty slack") to keep warm

3. There was no wind for four days
4. A layer of cold air was trapped under a layer of warm air – this is known as a temperature inversion
5. Air was full of smoke particles
6. Exhaust fumes from vehicles and smoke from steam locomotives

The United Kingdom's Meteorological Office calculated that the following chemicals were emitted **EVERY DAY** during the London smog:

1,000 tons of smoke particles	140 tons of hydrochloric acid
2,000 tons of carbon dioxide	14 tons of fluorine compounds

370 tons of sulphur dioxide which was converted to 800 tons of sulphuric acid

The smog created all kinds of problems for the people of London.

1. Visibility was extremely low – people could not see their feet!
2. Clothes became covered in smoke particles
3. Faces, nostrils and snot were black in colour.
4. Smoke particles collected in the lungs. The chemicals in the smog irritated the linings of the lungs and people suffered from breathing problems, coughs and bronchitis. In the five days of the London smog, 4000 people died. 100,000 people were ill and had difficulty in breathing.
5. The smog entered houses and buildings. Cinemas were closed – people could not see the screen!
6. Sporting fixtures were cancelled.
7. People got lost and could not find their way home.

8. Cars were abandoned. Drivers could not see to drive. Ambulances could not find their way to hospitals. Milkmen could not deliver milk to their customers.
9. There was no movement of ships, boats and barges on the River Thames.
10. Hospitals were full of patients and morgues were overflowing with dead bodies.
11. There was an increase in crime, especially burglary – it was easy to escape capture.

It was reported that cattle in Smithfield's cattle market choked to death. Some cattle had to be put down to stop them suffering.

On the 9th December a westerly wind started and blew away the smog. Deaths occurred after the smog had disappeared and it is thought that a total of 12,000 people died because of the London smog.

Because of the London smog and its effects, the government passed in 1956 the Clean Air Act which introduced smoke free areas across London and the banning of burning coal and increasing the production of smokeless fuels.

Day time photograph showing the difficult driving conditions in the London smog 1952.

Day time photograph of the London smog.

A daytime photograph of a man with a torch showing a truck driver the way through the streets of London during the smog of December 1952.

A man with a burning torch is leading the bus driver through the streets on London during the smog of 1952.

Websites

Chapter 1

commons.wikimedia.org/wiki/File:cat_with_mouse.jpg

commons.wikimedia.org/wiki/File:cat_eating_a_rabbit.

https://archive.org/details/extinctbirdsatte00roth

https://commons.wikimedia.org/wiki/File:Emerald Hours in New Zealand (1906) · Lowth · 200.jpg

Chapter 2

commons.wikimedia.org/wiki/File:Tree Sparrow August 2007 Osaka Japan.jpg

china-underground.com/wp-content/uploads/2022/05/sparrow-2.jpg

Chapter 3

https://commons.wikimedia.org?wiki/File:starling_(5503763150).jpg

https://commons Wikimedia.org/wiki/File:A_murmuration_of_starlings_at_Rigg_-_geograph.org.uk_-_4302954.jpg

medium.com

https://avitrol.com/pages/avitrol-for-starlings.html

Chapter4

commons.wikimedia.org/File:Brown tree snake aerial bait cartridges.jpg

commons Wikimedia.org/File: Micronesian Kingfisher 1644.jpg[File:Endangered Guam Rail commons Wikimedia.org/File:Endangered Guam Rail Guam DoA commons.wikimedia.org/File:Brown tree snake (Boiga irregularis) (8387580552).jpg.

commons.wikimedia.org//File:USDA dogs sniff out snakes 150430-F-EP111-265.jpg Chapter

Chapter 5

https://military history.fandom.com/wiki/HMS_Triumph_(1764)?file=HMS_Triumph_1764.jpg

https://en.wikipedia.org/wiki/Minamata_disease#/media/File:Tomokos_hand.gif

f620f53e3c56f043a867360e82b93b26

jsis.washington.edu

Chapter 6

b039592d-b7a5-4bef-8cd2-29ed9f742b77

b9c3586b-8623-4230-8113-11db90b8c909-22797

Inferno-Down-Under-600x600

FlannelFlowerKatoombaStraliaWeb_296

R (5)

Vombatus_ursinus_-Maria_Island_National_Park

ac-offplat-oz-firesBurned-out-car-surrounded-by-gum-trees-in-the-aftermath-of-the-Australian-wildfires

Chapter 7

Damage-ex-1-768x644

OIP (1)

Reynoutria_japonica._2020-08-30,_Seldom_Seen,_03

Reynoutria_japonica_in_Brastad_1v3-78c

The-root-system-of-the-Japanese-knotweed-through-the-seasons-occupies-metres-underground-1536x1152

Chapter 8

R (4)

R (3)

Untitled-design-2025-01-02T214643.653

India_Today_Archive-F149SI11-001336_IT_1575314130292

Chapter 9

Jamaican_Petrel

Jamaican+Giant+Galliwasp+-+Simon+J.+Tonge

Rattus_norvegicus_-_Brown_rat_03

BC_ZSM_Lep_73360+1425055428

Cattle-Fever-Tick-2

640px-Mongoose_5_(32151641192)

Chapter 10

024182300116-1

__opt__aboutcom__coeus__resources__content_migration__mnn__images__2015__06__kudzu-monsters-11-5ec0c3733087407cb8d223ac65090937

ZZBIhx6OtToJlLujQGAC_LPFZfigD0XfMCxYXYXugwA

Dust_storm_approaching_Stratford,_Texas

Maxresdefault

The_Dust_Bowl

ingredients-as-old-as-st-nick

Kudzu_filephoto

Chapter 11
88efda3cf3
100421-G-XXXXL-_003_-_Deepwater_Horizon_fire
211005124857-01-wildlife-oil-spill-file-super-169
1-turtlerescue-large
https://earthsystemsscience oilspill.weebly.com/
uploads/7/7/3/9/77399346/1978602.jpeg?503
bpoilspill

Chapter 12
Contaminated soil under the north track
derailment-3-ap-rc-230206_1675690785810_
hpEmbed_4x3_992
Untitled-design-2025-01-02T214643.653
OIP (3)
home-cleaning
ohio-train-crash

Chapter 13
749px-Emu_1_-_Tidbinbilla
emu-close-up_2022-09-23-152909_touvemus
Dromaius_novaehollandiae_-zoo_-two_feet
Deceased_emu_during_Emu_War
Lewis_Gun_during_Emu_War
emu_state_barrier_fence_emu_wars

Chapter 14
Valdez_Trash_Pile.jpg
exxon-valdez-docked
8233e0e24e8eb29877_2-exxon-valdez-rock
0l6rjtbzk01drbfxl6ds-1024x680jpg
MR 2001 P0000167A

Chapter 15

pngtree-aral-sea-catastrophe-sandy-salt-desert-on-the-place-of-former-picture-image_2295393

1412075992336

Aral_sea_1985_from_STS(1).jpg

689px-Stamp_of_Kyrgyzstan_115-119.jpg

Aral_amo_2008120_lrg.jpg

aral-sea-dying.jpg

R(8).jpg

Chapter 16

800px-Shelterbelt_Project_planting_areas

01_jackrabbit_opener.webp

640px-Dust-storm-Texas-1935.png

21171.12.jpg

The+Dust+Bowl+The+Dust+Bowl+refers+to+a+period+of+severe+dust+storms+and+soil+erosion+in+the+Great+Plains+during+the+1930s.

82cf78476ceeb400fd6aafb73b5050cf.jpg

The_Dust_Bowl

IH109310

Dust_storm_in_Spearman,Texas,_Wea01422

dust-bowl-016

Chapter 17

p07wnxrp

OIP (8)

OR2e8oY-scaled

kew-bus-conductor-carrying-a-flare-to-guide-driver-flickr-alan-farrow

www.ingramcontent.com/pod-product-compliance
Lightning Source LLC
Chambersburg PA
CBHW041219070526
44584CB00001B/13